Quantum Physics Made Simple

The Introduction Guide In Plain, Simple English For Beginners Who Flunked Maths And Science

© Copyright 2020 - All rights reserved.

The content contained within this book may not be reproduced, duplicated nor transmitted without direct written permission from the author or the publisher.

Under no circumstances will any blame or legal responsibility be held against the publisher, or author, for any damages, reparation, or monetary loss due to the information contained within this book, either directly or indirectly.

Legal Notice:

This book is copyright protected. It is only for personal use. You cannot amend, distribute, sell, use, quote or paraphrase any part, or the content within this book, without the consent of the author or publisher.

Disclaimer Notice:

Please note the information contained within this document is for educational and entertainment purposes only. All effort has been executed to present accurate, up-to-date, reliable, complete information. No warranties of any kind are declared or implied. Readers acknowledge that the author is not engaged in the rendering of legal, financial, medical or professional advice. The content within this book has been derived from various sources. Please consult a licensed professional before attempting any techniques outlined in this book.

By reading this document, the reader agrees that under no circumstances is the author responsible for any losses, direct or indirect, that are incurred as a result of the use of the information contained within this document, including, but not limited to: Errors, omissions, or inaccuracies.

Contents

Introduction 5

Chapter 1: Quantum Physics - The Basics.......16

   What Is Quantum Physics? 18

   Brief History of Quantum Physics 24

   The Standard Model of Elementary Particles 32

      Fermions 33

      Bosons 34

   What Is a Quantum Leap? 37

   Particles Behaving Like Waves: What's This All About? 39

   The Act of Measurement 44

Chapter 2: Quantum Physics - The Fascination 49

   The Superposition Principle 50

   Schrödinger's Cat..53

   Time Traveling: From Electrons to... Everything 59

   The Many-Worlds Theory in Quantum Physics 64

Quantum Entanglement: Explaining Destiny? 69

What Is Quantum Tunneling? 74

The Future of Quantum Mechanics: The Quantum Computer and Its Applications ..80

The Future of Quantum Mechanics: Teleportation? 84

Chapter 3: Quantum Physics - The Battle 87

Quantum Mechanics and General Relativity Incompatibility 90

Einstein's Explanation vs Quantum Mechanics 94

Why Do We Accept Quantum Mechanics? .97

Conclusion 106

References 116

Introduction

There is a world of sheer fascination out there. A world of underwater creatures we have no name for, of microscopic living beings that have survived every single catastrophe on Earth, of stars that have died out thousands of years ago, and yet still continue to beam their light on us.

Mankind has grown a lot. From the very first spark of fire to the fast-moving Internet connections that have brought billions of people together in a cyber home we are still learning to understand and play with, we have walked a path of thousands and thousands of years.

What might be shocking to some of you, our dear readers, is that this is just the beginning of it all. If we were to compare the evolution of mankind so far, to the development of a child, we would still be in our very early infancy.

We see the world around us and we perceive it with all our senses. We have learned to coo as a sign of communication and intelligence. We have learned to reach out with our hands for food and information.

And yet, we still have a very long way to go. We still have to learn how to walk and talk, how to grow, how to reach the stars, and how to embrace the beauty of everything around us beyond the limitations of our own cradle.

We owe our evolution to a bunch of pretty amazing people from a very wide range of fields of knowledge - artists and writers who knew how to transgress meanings and create a language that moves past borders, nations, and traditions to encompass human emotion in all its glory. Scientists who knew how to ask the right questions and how to give answers that were never finite, but always left room for *more knowledge* to be acquired.

The men and women who shaped this world are, without a doubt, an exceptional bunch. They are more than just thinkers and doers - they are shapers of the world in which we have all learned to live.

More than anything, though, they were visionaries who knew there was something *more* to what they did, every single time, who knew that self-sufficiency will lead us nowhere, but childlike curiosity and experimentation will lead us everywhere.

Literature, music, arts, physics, chemistry, mathematics, history, geography, geology, computing - the garden of human knowledge has grown to be more than impressive. And in this vast garden, a tree has sprouted with big promises for the future: Quantum physics.

For the vast majority of people, quantum physics is pretty much synonymous with (if not worse than) rocket science: An amalgam of sometimes contradictory theories that don't seem to have much of a *real-life* application.

After all, you know chemistry can help you dye your hair and get better medication. You know biology helps you understand your body and that it lies at the foundation of better healthcare. And you know that classic mechanics helps your car move from point A to point B.

But - quantum mechanics?

It seems to be one of those things that circles the high spheres of academia with no chance of touching ground very soon. Right?

As it happens, quantum physics holds the key to our evolution.

Want better solar energy?

Quantum physics might just hold the answer.

Ever dreamed of going into space and discovering new planets?

Quantum physics might be the *thing* that gets us there, eventually.

Always wanted a robot to do your house chores and to have intelligent discussions with over dinner?

You know the answer by now: Quantum physics.

At the moment, most quantum physics revolves around theories, sure. But once those theories are properly laid out and set in stone, we will move into a whole new era: One where information can flow freely and safely, one where we finally understand where we come from and where we are going, and one where even your wildest dreams will become reality.

The book at hand is by no means attempting to explain *everything* there is in the field of quantum physics: It would be foolish to think that a few pages in a book will actually be enough to encompass the grandeur of such a topic.

What we do aim for, however, is to create a book that will help you understand the basics in quantum physics - and even more than that, that it will stir your curiosity to learn more. The information is out there in the open - all you have to do is reach out and absorb it.

Why would you even do that?

Because quantum physics is fascinating, because it might finally give us the answers we have been looking for ever since we became sentient beings, and, because, even as a "layman", watching the mysteries of quantum physics unfold before our eyes is a spectacular show to behold.

We invite you to be part of this show - the one that will shape the future of mankind.

Together, we will move through three main chapters. The first one will explore the very basics of quantum physics: The definition, the history, the Standard Model of Elements, quantum leaps, wave-particle duality, and, finally, the measurement problem in quantum mechanics.

If the first chapter will be a bit heavier in theory, the second chapter will show you why exactly quantum physics is so wildly exciting. Throughout this chapter, we will explore some of the most interesting applications of quantum physics theories and how they might be closer to realization than you may even imagine.

Last, but not least, we will also explore the relationship between quantum physics and traditional physics, as well as how the two might be able to survive together in the future, to create a whole new perspective on... Everything we know.

From a branch of physics that was regarded as fringe (if not proper crazy) to a branch of physics that has grown to be an ambitious goal for some of the brightest minds of our century (Stephen Hawking and Michio Kaku are just two of the names we can drop here), quantum mechanics has changed a lot.

And yet, down at its very core, it still deals with some of the most perplexing paradoxes we have ever had to face as humans. Deep down, quantum physics is about advancement like never before - about breaking the barriers of knowledge not only in theoretical physics, but in every single area of development in the history of mankind.

Engineering, IT tech, communications, transportation, time traveling, the secrets of the universe and how we can actually travel through it, destiny and even the final reconciliation between religion and science - they all have room in quantum theory because, like it or not, this is the *it* thing happening in science right now.

Yes, crypto currency has taken over the world by proposing the decentralization of information. And yes, the Big Data flow is bigger than ever (as such, we need tools bigger than ever to be able to manage everything).

But if you go really deep down to the bottom of all these advances, you will learn that quantum physics wears the crown - a heavy one that seems to be inexplicably intricate and paradoxical at the same time, but, perhaps, would be the most precious crown human advancement has ever worn.

Quantum physics is not about empty theories. It is not theoretical gibberish that translates into nothing. Borderline between philosophy and science, quantum mechanics has grown to be the engine of our movement onwards as mankind.

Regardless of who you are and what you do for a living, you will find that quantum physics can help you transpire the boundaries of knowledge today. That it can help you be excited about science in general, and about the future in particular. That it can affect you very directly in ways that you may not have considered before.

For this reason, we have pulled together a book that doesn't aim, in any way, to encompass everything that quantum mechanics represents, but a book that aims to put it all in a nutshell - so that people like you can have a basic idea of what is going on in science and how it might affect you sooner, rather than later.

For many people, the realities of quantum discoveries have long been a far-fetched dream. But as you will see in the book, genuine advances are being made every year. In 2019 alone, scientists have broken through small experiments that have been proving what others could only postulate in theory.

It is thus quite easy to imagine how the future will come faster than ever. Teleportation and time travel are no longer the material of fiction, but the goals towards which real scientists in massively important institutions of the world are aiming.

We hope this book will help you understand the foundations of quantum physics in the simplest and most exciting way. We hope the pages ahead will help answer your questions - and inspire you to ask new ones. More than anything, we hope this book will give you *hope* for the future. There is a world of opportunity out there knocking on our door - and quantum physics might just be the discipline to hold the pass code to opening the gates of the bright future we have always imagined.

We have grown a lot - us human beings. We barely knew how to make our own fire just a million years ago (potentially) - and here we are now, dreaming of the stars and of the ultimate knowledge.

Are you ready to become a *scientific* dreamer, too?

Chapter 1: Quantum Physics - The Basics

For the vast majority of people, the term "quantum physics" is closer to "rocket science" than it is to "the wonders of the universe". And that's a real pity, actually. Most of you might think of boring formulae and explanations when thinking of physics - but the real truth is that both "traditional" physics and quantum physics are pretty much the sciences that hold the secrets to the universe: The *whys* and the *hows* of the way in which the entire cosmos works.

No matter who you are and what you do for a living, quantum physics will bring a whole new perspective into your life on so many things that it is absolutely impossible to ignore it. How could you, when you know that quantum physics is at the foundation of what you are, in the background of your fate spinning your life, and at the core of your very way of "functioning" as an intelligent being of the universe?

Almost borderline between science and spirituality, quantum physics might finally be able to explain the unexplainable and bring us closer to the essence of the world and help us transgress the borders of thinking that have been limiting us thus far.

This chapter is all about helping you discover quantum physics at its very basics: What it is, where it comes from, and the basic theories that define this science. We invite you to discover, step by step, the beauty of a discipline of study that has been long considered to be a mystery and an insurmountable topic at the same time.

Let's dive in and uncover the basics of quantum physics!

What Is Quantum Physics?

To understand what quantum physics is, you must first go to the "mothership", i.e., *physics*. For a lot of people, *physics* is that boring subject in school you have to be a real "nerd" to like: The one that is even worse than mathematics and even more difficult to understand than chemistry.

For many other people, physics revolves around mechanics, or, in layman terms, "how cars work". While it is definitely true that physics deals, among many other things, with how cars work, it is also worth noting that mechanics is only one branch of physics, and that it also deals with a bit more than just how cars work, but lays at the very foundation of car-making, alongside with electronics - another branch of physics.

The etymology of the word "physics" is pretty fascinating: It comes from the Greek word *physike,* which used to mean "knowledge of nature". As such, the definition of physics is tightly connected to *nature* and getting to know it. Many define physics as a *natural* science that studies matter, how it behaves through space and time, and how it connects to energies and forces.

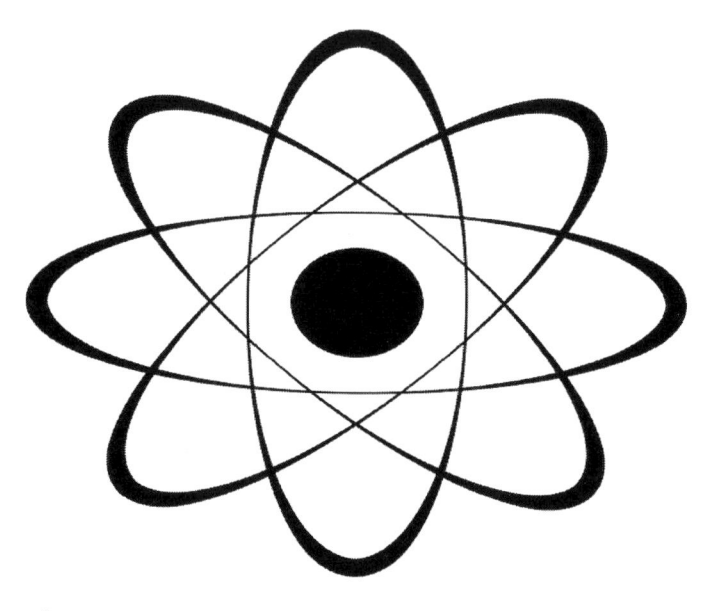

There are no less than nine branches of physics:

- Nuclear physics
- Atomic physics
- Geophysics
- Biophysics
- Mechanics
- Acoustics

- Optics
- Thermodynamics
- Astrophysics

Each of these deals with a different aspect of matter and different types of matter (such as in the case of nuclear physics, for example, which studies the way in which *nuclear* matter behaves in different contexts).

In addition to this categorization, you may also find people talking about *classical physics* and *modern physics* as well, which is a way to look at this natural science from the perspective of its evolution in time.

So where does quantum physics lie in this entire paradigm, you may ask?

Well, you see, quantum physics is a bit of an odd "animal", because it is sometimes used as a synonym to modern physics - so it simply comes as both a continuation of traditional physics and as an antagonist as well. Although most modern physics revolves around quantum theory, it is worth noting that, at large, it is still considered to be only a direction of modern physics.

To better understand the relationship between modern physics and quantum physics, consider the fact that the beginning of modern physics was marked by two major theories and theoreticians:

- Max Planck, who dealt in quantum theories (Planck's Constant is the most famous one, and it entailed that the energy and frequency of light is proportional, which later on led Einstein to postulate the fact that light exists in small quantities of energy called "photons").
- Albert Einstein, whose major work was related to the theory of relativity and the photoelectric effect. The first postulates, in short, that massive objects can cause a distortion in space and time, which is felt as gravity. The latter says that light does not exist in waves, but in quanta (small pockets of energy), as we have also mentioned above.

If "big" physics deals with things you can more or less see (or at least perceive, especially if you run experiments), quantum physics deals in the *very small* parts of matter. In fact, this is its actual definition as well: The science that deals with the atomic and subatomic levels of matter.

That might not sound like much. However, quantum physics has gone so in-depth (and continues to do so), that it might actually be the one to finally explain *everything* in the universe.

Everything we've never known. All the questions we always sought an answer to - the very essence of *life*.

Can you think of quantum physics as a boring topic when you look at it through this perspective? When you know that it is the science that will finally help us understand so much more about ourselves, our place in the universe, and where all of this is leading?

As you will see later on in the book, the world of quantum physics is exciting and fascinating in every sense there is. Not only does it appear to hold the key to all the things we never managed to achieve (like teleportation, or understanding fate and destiny, for example), but it is also a contradiction to a lot of very well-established theories, including that of general relativity brought forward by Albert Einstein himself.

There is a war for knowledge out there, and quantum physics just "happens" to lie at the very core of it.

We bet we made you curious!

Brief History of Quantum Physics

Before we dive deeper into the details of how quantum physics came to be, it is important to eliminate confusion related to quantum physics and quantum mechanics. The reason we believe this to be important is because you will see both terms used across the history of quantum physics and in contemporary talks, papers, and articles.

To clear things up: Quantum physics and quantum mechanics are two terms that are very frequently used interchangeably. Indeed, the difference between the two is minor and most scientists agree to use them as near synonyms, rather than different categorizations of the natural science called "physics" (or different branches of quantum sciences in general).

Of course, there are several examples of when quantum physics and quantum mechanics are not entirely interchangeable. For example, if you discuss quantum physics and quantum chemistry, you will refer to the first using the "physics" term, rather than the "mechanics" one, so as to avoid any kind of confusion.

Other than that, the two terms mean quite the same: The study of very small particles in matter. Throughout the remainder of this book, we will use them interchangeably as well, as per the model already used by the scientific community of today.

The history of quantum physics (or quantum mechanics, if you want to call it this way) draws back to the beginning of the 20th century. To understand the context in which this new branch of physics lies, you have to know that, the end of the 19th century and the beginning of the 20th came with a complete revolution in multiple areas of knowledge.

Darwin's Theory of the Species was already widely accepted. Nietzsche declared God to be dead. And the very perception of human life had started to drastically change. From the linear view of the previous centuries, humans were now seeing themselves through a broken mirror: The rules of the past were being torn apart, and a new, modern human being was arising from the ashes of the previous centuries.

It is this human being that started the two atrocious world wars. But it is also this human being that went to the Moon, invented computers, and connected billions of people through the invisible (yet incredibly powerful) means of communication we have grown to refer to as *the Internet*.

In this context, science was at a high level of unrest. The theories that had worked until then stopped being perceived as *ultimate*. Mankind needed a new approach on science in general, and that meant that some of the past theories had to be left aside, to allow room for new theories that would push mankind forward.

This is how quantum physics was born: As a response to all the questions traditional physics simply couldn't answer any longer. At first, the realm of quantum physics was disparate: One theory here, another one there, slowly coming together to shape not only a completely new branch of physics, but to mirror a completely new view on life itself.

As it frequently happens, quantum physics was not born out of sheer nothingness. Nobody woke up one day to say: *This is it, traditional physics doesn't cut it anymore.*

In fact, the terrain for modern physics was prepared centuries in advance. More specifically, the very roots of modern quantum mechanics traces as far back as the 17th and 18th centuries, when physicists such as Robert Hooke, Christian Huygens, and Leonard Euler brought forward a new theory of light, which was meant to be seen as *rays*, rather than colored particles (which is how Sir Isaac

Newton described light before this) (PhotonTerrace, 2020).

Things started to change when Faraday discovered cathode rays in 1838 (Wikipedia, 2020). His studies were continued by Gustav Kirchoff and Ludwig Boltzmann, whose theories suggested that light (and energy states of a physical system in general) can be discrete.

In 1900, the hypothesis of Max Planck came into the world to say that energy should be understood in terms of discrete quanta (energy packets). This marked the beginning of modern physics. Five years later, Albert Einstein brought forward the theory of the photoelectric effect.

In science, things rarely come as a state of fact per se. Although atoms had been accepted for nearly a decade after 1905, photons were still widely regarded as an unaccepted concept. In fact, even big names associated with the development of the quantum theory rejected photons (including, but not limited to Niels Bohr, who developed the theory of atomic structure).

Bohr's theories were continued by Peter Debye and Arnold Sommerfeld, who both extended the theory of atomic structure by adding elliptical orbits into the theory. This represented the foundation of the old quantum theory, which was also the point from which atomic physics was developed as well.

By the mid-1920s, Bohr and Heisenberg closed the old quantum theory and started to call what was previously referred to as "light quanta" photons.

In the 1930s, quantum mechanics was pushed forward by names like David Hilbert, Paul Dirac, and John von Neumann. It was the first time the quantum theory was leaning towards the statistics nature of nature as we know it - and also the first time the "observer" was brought into discussion (only at a philosophical level at this point).

During the 1930s, quantum physics branched into several sub-disciplines, including quantum chemistry, quantum optics, quantum information science, and quantum electronics.

You see, physics is not a field that *happened* and we are now just studying it in textbooks. It is a field that is *happening* - an exciting one worth following if not for anything else, then at least for mere curiosity about the world in which we live and its place in the grander scheme of the universe.

From the 1940s onwards, quantum mechanics evolved into massive theories that are still being worked on - like the Quantum Gravity theory and the String Theory, for example.

In time, a very big differentiator started to delineate traditional physics from quantum physics. If in classical physics all the properties of objects had to be calculated, quantum mechanics takes a more probabilistic approach of nature. In quantum physics, energy can be either matter or wave, depending on certain characteristics. This central trait of quantum mechanics opens a whole new world of opportunities for practical applications, as you will see later on in this book.

What we will approach in this chapter (as well as the following ones) are some of the most interesting theories in quantum physics. Some of these theories have been proven, while others are still in the works - but as you will see, all of them are related to our actual evolution as a species and how we plan on "doing things" from here on.

This is precisely why quantum physics is so interesting, actually: It is happening right now and, in one way or another, it will determine our very future.

If that's not exciting, we don't know what is!

The Standard Model of Elementary Particles

If chemistry has its *Periodic Table of Elements*, quantum physics has its Standard Model of Elementary Particles - a set of formulae and measurements that describe elementary particles and the way they interact. Where the Periodic Table of Elements categorizes atoms based on their specific characteristics, the Standard Model categorizes elementary particles in two main groups: Fermions and bosons.

Developed at the beginning of the 1970s, the Standard Model came as a means by which quantum physicists attempted to standardize what was already known about particles and forces. This effort had the goal of furthering research in the field and helping other scientists refer to a standard measurement of the elementary particles, rather than to a disparate cumulus of theories and theses.

In addition to presenting what was already known to the scientific community (and, as such, what was already accepted by it), the Standard model also helped predict the existence of particles that were not discovered yet (such as the (in)famous Higgs boson).

The Standard Model is considered to be one of the best formulated theories in particle physics. However, it still has gaps scientists are trying to uncover - like how general relativity's approach on gravity can be integrated into quantum theory, or to give an explanation on why there is more matter than antimatter in the universe.

Fermions

According to the Standard Model, fundamental particles are categorized into groups that are related to each other according to certain rules. For instance, two fermions cannot be in the same place at once. This rule allows them to build, block upon block, *everything* (from atoms, to the shoes you're wearing, and the planets of the universe).

There are two categories of fermions:

- Quarks - which can combine into protons and neutrons. For instance, a proton consists of two up quarks and one down quark, which are connected by strong nuclear force.
- Leptons - which include electrons, muons, taus, as well as neutrino electrons, neutrino muons, and neutrino taus. These neutrino particles are very interesting because they are barely noticeable, but they might still have a heavy influence on the world.

Bosons

If fermions make up the world, bosons act as liaisons between different fermions. They are, in some ways, mediators between the forces that act on matter by either binding or repelling it.

Bosons explain why we can't do certain things - such as why light comes in the colors of the rainbow, why human beings can't walk through walls, and so on.

Breaking the mystery of bosons would be a tremendous achievement for mankind. We might not be daring enough to think of X-Men-like superpowers that will allow us to walk through walls (to circle back to the aforementioned example), but managing to control certain bosons would definitely give us *power* that would have been considered superhuman just a couple of hundreds of years ago.

In the bigger family of bosons are included:

- Photons (which communicate the electromagnetic force)
- Gluons (which deal with the strong nuclear force needed to bind quarks together and create protons and neutrons)
- W & Z bosons (which have to do with the weak nuclear force)

- Higgs boson (which is meant to explain why some particles have mass when certain conditions are met).

Of course, this is all a very short version of the entire story. As you can imagine, the information in the Standard Model of Elements lays at the foundation of future science.

For instance, do you remember how years ago everyone went hysterical about a scientific program in Switzerland that was risking creating a "black hole" that would absorb us all? What they were actually trying to do was to isolate the Higgs boson, which was a beginning in the understanding of an array of concepts, including (but not limited to) dark matter.

Fascinating, right? And this is just a very, very small piece of the entire ensemble of quantum physics.

What Is a Quantum Leap?

Aside from the infamous Higgs boson, "quantum leap" is a term that has been so commonly used in mainstream media and knowledge that it is nearly impossible not to have heard about it if you have ever watched a science fiction movie (or anything to do with science, really). There is even a TV show from the late 1980s that carries the same name!

In fact, the quantum leap concept has been so widely used that it has even entered general vocabulary, in the sense of a "major, drastic, and sudden change".

For centuries, people have not quite understood what quantum leaps do, how they function, or what applications they might have. Recently, though, it has been discovered that quantum leaps are quite the trick: They are not as sudden as we might have thought, and they are quite gradual. It's just that we cannot actually *see* them, because they happen in about four microseconds. Sure, for the human perception and for the human mind, that's nearly as "good" as being instantaneous - but in the world of physics, that is not *instantaneous* as per the definition (Ball, 2019). But what is a quantum leap in physics, in the end?

Well, a quantum leap is the change that occurs to an electron from one energy level to another. Scientists have *guessed* that these leaps happen, but until recently, only scarce experiments have been able to prove it.

The discovery that quantum leaps are deterministic, reversible, continuous, and coherent (made in 2019) is actually extremely important because it might change the entire paradigm of the standard interpretation in Quantum Mechanics.

As for the applications of this discovery, suffice to say that it will help scientists push forward quantum computing (which, as you will see later on in the book, is quite important for the evolution of mankind).

Particles Behaving Like Waves: What's This All About?

The relationship between physics and waves might sound strange to you if you haven't had much tangency with the physics theories developed throughout the past century or so.

However, waves have grown to be a core concept of the realm of physics - so much so that pretty much every standardized test that includes physics within its topics (like HESI, for example, a test you have to pass to become a medical nurse) includes waves within its curricula.

In quantum physics, it is assumed that light (a form of energy) behaves as both particles and as waves, depending on certain circumstances. This is, as we have mentioned before, the very core difference between modern and classical physics.

When scientists talk about particles that behave like waves, they frequently refer to this concept as "wave-particle duality". This phenomenon is a relatively new concept in the world of physics, and it comes to overthrow classical concepts of both "particles" and "waves", which are now considered to be unable to fully describe the behavior of objects on the quantum scale.

In other words, in classical physics, scientists referred to particles and waves as two different concepts. In quantum physics, however, scientists like Max Planck, Einstein, and Niels Bohr (just three of the forerunning names in the incipient stages of quantum mechanics) have concluded that particles frequently exhibit a wave nature and that waves can exhibit a particle nature as well.

They are not two completely delimited notions - they dance together to create the world in which we live (and everything outside of it, which we have yet to discover and know).

As we have also mentioned before, quantum mechanics is a pretty young field, and actual research is being done as we speak. One of the major pain points in quantum physics is precisely explaining the meaning of the wave-particle duality.

The very views on wave-particle duality have evolved quite a lot in time. If Bohr saw it as a metaphysical fact of nature (and considered the entire concept as an aspect of complementarity), Heinseberg reconciled the concept of causality to that of the wave-particle duality.

Later on, it was postulated that all the information about a particle is embedded in its wave function, which connects, once again, to the idea according to which particles are waves and which are, well, particles.

And now, let's explain all of this in plain English.

A particle is something noticeable, finite. You can pinpoint it (maybe not by just looking at it, but with the right tools, you can *see* a particle and point your finger at it, at least imaginarily).

A wave, on the other hand, is not finite. It is everywhere. It wiggles. It moves around and covers the entire space-time continuum. Compare this, if you wish, to the waves of the ocean. You cannot actually say that a single point in space or time is a wave - rather than that, the movement of the water creates endless waves that move back and forth.

In quantum physics, light (and matter in general) is considered to be both a particle and a wave. When you look at it and pinpoint it, light (and matter, again) is a particle. When you analyze the probabilities light (or matter) could have (e.g. its actual probable measurements), you are analyzing it as a wave.

What does this have to do with real life?

Well, everything - or at least a *lot*.

For instance, a better understanding of the wave-particle duality will help us build better photovoltaic panels (solar panels, if you want to call them this way). When we understand that electromagnetic radiation has the behavior of a particle, rather than a wave, we also understand that it can affect electrons that are free from their atoms. In exchange, this creates energy (which we can use to heat our homes, run our computers, and so on).

MRI imaging, quantum computers, microchips, cosmology, and even the super science fiction concept of teleportation - they all trace back to the fact that particles behave like waves (and waves behave like particles).

The Act of Measurement

The act of measurement (otherwise known as the measurement problem) is an issue in quantum mechanics, and it is related to how wave function collapse occurs. A wave collapse, in its turn, is the concept used to describe the moment a wave function changes from several eigenstates to one, due to the influence of the external world. Because such a collapse cannot actually be observed, several interpretations have arisen to explain the phenomenon.

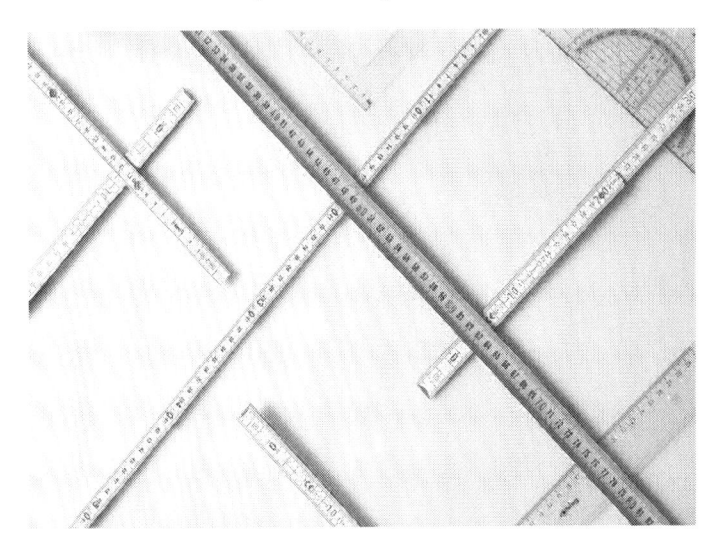

More specifically, these are some of the main interpretations that have been given to the act of measurement:

- The Copenhagen interpretation is one of the oldest theories (and one of those which are still common in quantum mechanics). According to this interpretation, the act of observation results in the collapse of the wave function. The interpretation fails, however, to explain *how* this happens (the "mechanism" behind it).
- The Hugh Everett Many-Worlds interpretation says that there is only one wave function (or the superposition of the universe). As such, the wave function never collapses, so there is basically no measurement problem. In this interpretation, however, the act of measurement is related to the interaction between quantum entities (such as the observer, the measuring instrument, and so on).
- The De Broglie-Bohm interpretation postulates that the information that describes a system includes not only the wave function, but also a trajectory that

sets the position of the particles. In this paradigm, the wave function generates the velocity field for the particles. According to this theory, when something interacts with the environment during the measurement procedure, the wave packets are separated in configuration space (which is why wave function appears to collapse, even if there is no actual collapse).

- The Ghirardi-Rimini-Weber interpretation proposes that wave function happens spontaneously. In this paradigm, particles show a non-zero probability of undergoing a wave function collapse (which happens once every hundred million years). And here's where things become interesting: The collapse itself may be rare, but the number of particles in the system make it highly probable that a collapse will occur somewhere in the system. The system is entangled by quantum entanglement, so the collapse of one particle will draw the collapse of the measurement apparatus as a whole.

- The quantum decoherence interpretation was set in the 80s, and some believe that it solves the measurement problem. Although it was initially brought forward in the context of the many-worlds interpretation, it has also expanded past that and some updates of the Copenhagen interpretation have adopted it as well. Instead of describing the collapse of the wave function, quantum decoherence explains the way in which quantum probabilities are converted into ordinary classical probabilities.

Although things have slowly started to settle down in the branch of quantum measurement, there is no unanimous theory to be accepted. On the one hand, this makes the entire field of quantum physics more confusing for beginners (and for those who work in research as well). On the other hand, however, it also makes it more exciting - because it means that the future might bring along new theories and new interpretations as well.

The act of measurement lies at the basis of the observer effect, which states that simply observing a phenomenon changes the phenomenon itself. This is connected to the idea of quantum entanglement as an explanation of *fate* or *destiny*, as you will read later on in the book as well.

From afar, all these very theoretical approaches to the nature of the world might seem just that - mere theory. But when you go in-depth, you will soon discover that solving these problems is not just a matter of academic achievement, but a matter of simply advancing mankind as a whole.

Time-traveling, parallel universes, destiny - they all seem like science fiction concepts. But to quantum physicists, they are as real as flying, electricity, or the Internet were once for the scientists of those times.

In the next chapter, we will take the time to explore some of these ideas and the quantum physics theories that lie at the foundation of what *might be* in the future.

Chapter 2: Quantum Physics - The Fascination

Quantum physics is more than just an exact science: It is a discipline that brings together philosophy, spirituality, and science. A discipline that might hold so many answers we have been looking for for so long that it is almost impossible to wrap our brains around the magnitude of the discoveries we're getting so close to.

This chapter is all about the most interesting discoveries and theories in quantum physics - those that will shape the world of the future with new technologies and new knowledge, and, as such, those that will push the entire world forward.

The Superposition Principle

The superposition principle is one of the most important concepts in quantum physics - one that needs to be properly understood because it lies at the foundation of many quantum applications.

According to the superposition principle, in all linear systems, the net response caused by two stimuli (or more) is consisted by the sum of the responses that would have been caused by each stimulus individually (Wikipedia, 2019).

That sounds a bit complicated, but what the superposition principle implies, in fact, is that, in quantum physics, a particle can exist in all states at once - so they can move at different speeds and have different energies all at the same time.

This happens because, as it has already been mentioned before, particles behave like both waves and particles at the same time. If you think of two ocean waves coming together, and you consider them as a collection of particles, you can either see them clashing and annihilating each other or coming together in a new state (i.e., a bigger wave), which is how traditional physics saw the interaction between two waves (and which is also the theory upon which numerous applications were built, including noise cancelling headphones, for example).

In quantum physics, the superposition principle feels rather counterintuitive for anyone coming from traditional physics, precisely because it assumes that a dynamic system can exist in all states, but that the "general state" is given by the overlapping of two or more states (where the "general state" is defined as what you actually see, or the state that is measured).

Basically, the superposition principle shows that everything is possible at the same time, and that the act of measurement (as it is preponderantly understood in the Copenhagen interpretation) is precisely what makes a certain "something" happen. Needless to say, this comes with a lot of real-life implications because it removes linearity from an equation.

One of the best real-life implications that comes with this is quantum computing, which relies a lot on the superposition principle. In traditional computing, humans and machines communicate with each other through a series of codes made up of zeros and ones. In this context, you can only work with zeros and ones alternatively. In quantum computing, however, due to the principle of superposition, you can work with zeroes and ones simultaneously.

The consequences of this are tremendous, and they are to be applied in a series of real-life technological improvements. For example, it is believed that quantum computing is one of the most important elements in developing advanced Artificial Intelligence (yes, like the one you see in the movies).

Schrödinger's Cat

Let's face it: The Internet is enamored with cats and everything they do. Every cat is famous these days - but in the world of science, there is no cat more famous than... Schrödinger's cat.

If you have heard of the term, it means you had at least a bit of tangential interaction with quantum physics. We hate to disappoint, though, and say that this concept has nothing to do with an actual cat (unlike in the famous example with Pavlov's dog, for example).

What "Schrödinger's cat" refers to, in fact, is a thought experiment (or a thinking exercise, if you want to call it this way) proposed by Schrödinger (and preceded by Albert Einstein).

In this paradoxical thought experiment, Schrödinger describes a cat that is locked in a steel chamber with a radioactive atom and a flask of poison that might be spilled by the radioactive atom. The cat may or may not die, depending on whether the radioactive atom is decayed or still emitting radiation. However, the state of the cat is not known until it is observed - so until someone opens the steel chamber to see if the cat is dead or alive, the cat exists in both states, both dead and alive.

Oddly enough, Schrödinger's cat example was meant to contradict the Copenhagen interpretation of the "act of measurement" and point out how ridiculous it seems to think that a photon or a particle can exist in all possible states until it is measured and collapses into a state.

Instead of surviving as a counterexample (and, by all initial intents and purposes, a *mock*) of the superposition principle proposed by quantum physics, Schrödinger's cat slowly started to be associated with it. Today, the cat example is given every time the superposition principle is explained by quantum physicists and experiments show that the superposition principle might just be true.

Furthermore, the famous cat (thought) experiment gave birth to new interpretations and explanations as well.

- In the Copenhagen interpretation, the cat becomes dead or alive when the box is opened.
- In the Many World interpretation, once the box is open, the cat is both dead and alive because the observer becomes entangled with the cat and splits into two states (an observer looking at a dead cat and one looking at a cat that is alive). However, because the two observers are

decoherent, they do not interact with each other.

- In the Relational Interpretation, every participant is an observer. For instance, the cat becomes an observer of the apparatus (the box and the atom with the flask of poison) and the human becomes an observer of the system (the cat and the apparatus). Until the box is opened, the two have different information about the reality, but when the box is opened, the two observers' realities "collapse" and become one.
- In the transactional interpretation, the apparatus (the box and the atom) sends an advanced wave back in time. At the same time, the source sends a wave forward in time. The two waves are combined, and the apparatus becomes an observer. The collapse of the wave function is atemporal, and it happens across the entire transaction between the source and the apparatus. As such, the cat is never in superposition - it is only in one state at a particular time, regardless of what the human does. In conclusion, the transactional

interpretation resolves the quantum paradox.
- In the Zeno effect interpretation, the environment becomes the so-called "observer". The human may peek into the box sooner or later, but the state of the cat is either delayed or accelerated due to its environment (i.e., the atom and the flask of poison).
- In the objective collapse interpretation, superpositions are destroyed instantaneously when a specific objective physical threshold (time, mass, etc.) is reached. As such, the cat dies or continues to live long before the box is opened and its state is observed.

Who would have known that a cat can generate so much debate in the scientific community? And yet, here we are. The famous Schrödinger's cat example is still being discussed to date, and there is no unanimous response to the problem. What remains a fact, however, is that quantum physicists all agree that the cat is (at least in a sense) both alive and dead until the state is observed. In other words, you, as an observer, will not know if the cat is alive or dead until you open the box to see it with your own eyes.

Understanding how to maneuver this situation is bound to attract a whole new series of practical applications in quantum physics - some of which are, as you will see, quite the science fiction scenarios.

Time Traveling: From Electrons to... Everything

Time traveling has long been a dream of mankind. We've written about it, made movies about it, and we've imagined it in every way there is.

What if we told you that, in quantum physics, time traveling is as much an object of research as new cancer cures are in medicine?

Time traveling may seem impossible to pretty much everyone else, but quantum physicists think they can solve the problem - first, at the level of the electron, and then, at the level of pretty much everything.

Until quite recently, most of the research made in the field of time traveling was run according to the general relativity principle (as explained by Einstein). However, quantum physics has brought a whole new perspective to light.

It all began with a Russian scientist called Igor Novikov, who postulated the self-consistency principle, which was supposed to solve the paradoxes in time travel. According to the general relativity principle, closed time-like curves would, in theory, allow time travel, but time travel paradoxes would be an issue (e.g. someone traveling from the present time to the past and making a change that affects present time itself).

However, according to the Novikov principle, if an event causes a paradox/ change in the past, the probability of that event happening (in the present, so that it can cause the past paradox/change) is zero. As such, time paradoxes cannot happen. So, for example, if you were to travel to before you were born, and accidentally made it so that your parents wouldn't meet, you would not exist in the present time, and, as such, you couldn't possibly travel to before you were born.

After the Novikov principle, quantum physics took two main directions when it comes to time traveling and applying the self-consistency principles.

- The Deutsch prescription (developed by David Deutsch)
- The Lloyd prescription (developed by Seth Lloyd)

The main problem in time traveling according to quantum physics is figuring out the time evolution equations for different density states that appear in the presence of time-like curves that are closed (also known as closed time-like curves or CTC).

How does this translate into plain English?

Basically, quantum physics has an issue with how time should be perceived. Most quantum scientists agree that, as per the quantum theory, time does not actually exist - it is a straight line in which all events happen at the same time (and yet, they don't, just like Schrödinger's cat is both alive and dead before the box is opened).

In theory, however, resolving the quantum problem means that we will all get a lot closer to time traveling because, if time does not exist and it is just a construct of the observer (us), then we can maneuver it as we want.

A major step has been made in this direction in 2019 when Markus Arndt and a team of scientists managed to observe a molecule in the state of superposition (Greene, 2019). Of course, they weren't able to *see* it per se, but were able to measure its different states. This is a major breakthrough, because it finally brought together classical and quantum physics.

More than that, it might mark the beginning of a whole new era in physics research, including in the field of time traveling. If we can observe superposition, we might be able to maneuver it at some point, and, as such, we might be able to move different large objects across the space-time continuum as we wish.

Only time will tell if we ever get to that point, but, at the moment, there is a very high chance that we do. The future looks brighter than ever - or should we say the present, past, and future are looking brighter than ever?

The Many-Worlds Theory in Quantum Physics

You have already been roughly acquainted with the Many-Worlds theory in quantum physics, as we have brushed over it both when explaining the measurement problem and when explaining the different views on the Schrödinger's cat mental exercise.

To circle back and provide you with the full picture of what the Many-Worlds theory entails, we have to define the Many-Worlds theory at large. Basically, what this branch of quantum mechanics says is that there are a number of worlds just like ours, but where the entireties of particles have different states.

In other words, the Many-Worlds theory explains the existence of the so-called "parallel universes" as you have probably seen in the movies. Basically, this theory says that there might be a large number of "you" versions in a number of worlds.

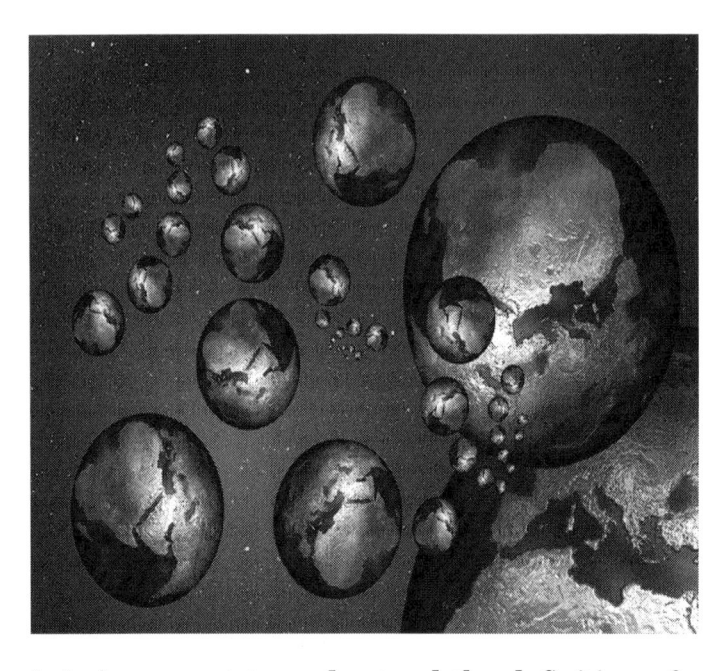

It is important to understand the definition of "world" in this context, as well as the definition of "I". A "world" is the totality of macroscopic objects (stars, planets, the space between them, and everything contained by them including, but not limited to human life).

"I", on the other hand, should be understood as a singularity. Let's say my name is John Doe, and there are millions of worlds with millions of John Does just like me whose lives have taken different paths according to the way in which the particles in those worlds have collapsed from superposition. If there are a number of John Does, however, there is only one "I", for "I" represents the specific John Doe in this world - here and now.

It all sounds very philosophical, but if you take a bit of time to analyze the problem, you will definitely understand where this goes. There are a large number of worlds in states of superposition, a large number of objects in states of superposition, and a large number of universes in states of superposition, making pretty much any scenario possible at the same time as the scenario in which you and I live right now.

As it has been shown in the previous sections of this book, the Many-Worlds theory has implications across the entire spectrum of quantum mechanics. The main issue with it, however, stems from the fact that superposition is not yet fully explained - and as such, the Many-Worlds theory could easily succumb should it be proven that superposition does not happen according to this paradigm.

On the other hand, if the Many-Worlds theory is proven, it means that everything about quantum physics is pretty much right, and that we do, indeed, live in just one of the near-infinite multitudes of worlds.

It is a thought-provoking exercise - and one that might actually push the boundaries of knowledge like nothing before. The very idea that there might be a huge number of worlds where things are so different and yet the same is mind-blowing:

A world where America never happened.

A world where slavery never happened.

A world where small countries are superpowers.

A world where Germany won the war.

And another world where Russia won the Cold War.

And from the multitude of worlds, you live in this one - which might actually be the best world there is, just like Voltaire's Candide kept repeating over the course of the eponymous bildungsroman of the 18th century.

It might just be that we also live in the worst of the worlds. Or that there might be a world out there where you never chose to read this book and that would be the only difference between *you* and the other John Doe.

The very thought of the Many-Worlds theory should help you understand why quantum physics is so important and so exciting at the same time - precisely because it might be the theory that makes all imagination seem wildly accurate and wildly incredulous at the same time.

Quantum Entanglement: Explaining Destiny?

For thousands and thousands of years, people have tied themselves to the idea of fate and destiny. You will find this concept in pretty much every religious system of the world. And more than that, you will find it discussed in philosophy, even outside of the religious paradigm and beyond the religious dogma.

In very short, what quantum entanglement says is that two particles relate to each other according to a pre-established path. They are forever tied to each other and will always follow the action that is "prescribed", in a manner of speech.

If you extrapolate this to macro-objects, it means that everything follows a prescribed path and that nothing is ever born out of an actual cause, but because the particles behind it have been pre-decided to act a certain way.

It is hard to imagine that everything we are and everything we will ever be is simply *written* and that we follow a prescribed path. The most direct and shocking implication of this is that we have absolutely zero free will - so pretty much every choice you have ever made was nothing but the simple following of a universal script that has already decided what you are going to do.

Destiny might seem like a concept that has nothing to do with science, but if you think of it like this, you need to get acquainted with the quantum entanglement theory. Far from religion and spiritual experiences, this theory postulates that we might as well just be pawns in a massive game of chess.

And then, if we are just following our fate without any kind of free will, who are the players behind the game of chess? Who is pulling the strings from behind the veil? Is it an entity? Is it a force? Is it something you and I and pretty much nobody else will ever be able to understand precisely because it is so big and hard to understand that we do not have the power to grasp the full meaning of it?

Would that be the final reconciliation between religion and science?

Quantum mechanics might be able to explain all this, but unlike in the case of the Many-Worlds theory and time traveling in the quantum point of view, quantum entanglement study results might scare us and it might have implications that are hard to fathom.

For instance, if someone becomes a serial killer and it is proven that they had absolutely no free will over their acts, is it still right to punish them? Weren't they just following a script, without even knowing that they are doing it?

Every good deed and every bad action we take would be easily excused. It is, after all, the fate you are following, right?

And if that is all true, where do we stand as human beings? We have known for quite some time that we are small in a very large scheme of things and that our entire planet is nothing but a grain of sand in an incomprehensibly large universe.

To many people (including scientists like Albert Einstein), this entire theory was as crazy as believing the cat in the box can be both dead and alive at the same time. It was but a mere figment of someone's imagination and nothing more - a provocative thought exercise, but nothing that would ever materialize into a real-life, proven scientific theory.

As it turns out, quantum entanglement is not that much of a fantasy theory. In 2019, scientists (Macdonald, 2019) have actually managed to capture it: Two particles inextricably linked to each other and forever affecting each other no matter how large the distance between them is.

But if the quantum entanglement theory proves to be true, it might show us that we are even smaller than we think we are.

And what is then left for us to do?

What Is Quantum Tunneling?

Before we close this chapter, we also want to dive into quantum tunneling, one of the most important concepts in quantum mechanics - and one that, just like quantum entanglement and superposition, might change our perspective on pretty much everything.

Quantum tunneling might also lie at the basis of cutting-edge discoveries in a lot of real-life fields, and it might explain a lot of questions scientists have posed about the universe itself.

To understand quantum tunneling, you must first understand the Uncertainty Principle as it was brought forward by Werner Heisenberg. According to this principle, we cannot measure exactly where a particle (electron, in his example) is in terms of location (position) or time (momentum).

While we might not be able to know where the particle is, we can model it using wave functions and give a very good, probabilistic approximation on where the said particle is. In theory, you can also give a probabilistic approximation of where the particle *will* be as well.

Let's start at the beginning, though, and circle back to quantum tunneling. In classical physics, when particles meet a barrier that they cannot surmount due to insufficient energy, they just don't move past it.

In quantum physics, however, scientists believe that particles can overcome the barrier even though they do not have sufficient energy to do so. The reason they are able to do this is because they behave like waves, and when they encounter a barrier, they do not necessarily end abruptly right then and there - the amplitude of the wave decreases, which would usually correspond to a drop in the probability that one (or more) of the particles in the wave moves past the barrier.

However, if the barrier is thin enough, then the amplitude might actually be a non-zero when it reaches the other side. As a consequence, it means that there is a finite possibility that some particles are able to actually tunnel through the barrier - which is precisely what quantum tunneling is.

To exemplify it in a simple way, imagine a ball (which represents the particle in our previously explained theory) that has to be pushed with enough energy to surmount a hill that stands before it. If the push is not strong enough, classical physics says, there is no way the ball can move over the hill.

In quantum physics, if you take the same ball (which represents the same particle) and place it in front of a barrier (let's say, a block of flats), there is still a probability that the ball will be able to surmount the barrier in front of it. The ball will stop behaving like a ball and it will start behaving like a wave.

To imagine this, think of light (which is a wave) that reflects and refracts from a surface. Our "ball" behaving like a wave will be reflected back, but some of the wave might transpire through the barrier. If there is a probability that the particle in the wave has made it through, then we can say that there is a small, but not non-zero probability that the particle has tunneled through the barrier.

Again, this might all sound like just theoretical gibberish, but the very fact remains that quantum tunneling is at the basis of very important discoveries. Some of the most common applications of quantum tunneling are flash drives. And some of the most daring ones imply the creation of compact fusion reactors that will generate more energy than ever.

Of all the "mad" theories quantum physics has ever come forward with, quantum tunneling might just be the most palpable one - and the one we are closest to precisely because bits and pieces of this principle are already applied in scientific research (in engineering, and not only in that, either).

The Future of Quantum Mechanics: The Quantum Computer and Its Applications

The last two sections of this chapter are dedicated to two of the most daring (and scary) practical applications of quantum mechanics: Quantum computing and teleportation.

Quantum computing is another subject we have briefly brushed upon in this book, but we would like to take a bit of a closer look at it now that we are towards the end, precisely because it might be one of the most palpable and immediate results of research done in theoretical quantum physics.

As we have also mentioned before, classic computing (the one we use now for pretty much every computer device we own) is based on a series of communications made up of "1"s and "0"s.

Quantum computing, however, would open the gates to more computing power by allowing us to communicate with machines in a language that brings together both ones and zeroes at the same time. In other words, a quantum computer can do faster calculations by using a smaller amount of energy - which means that there is an almost infinite amount of power to come out of a quantum computer (as compared to a classic one).

The immediate applications of quantum computing are both scary and exciting at the same time. One of the major breakthroughs that would happen immediately after the release of the first quantum computer that is actually functional at its full power is the way in which machines will be able to break certain codes. Clearly, this poses a major security risk for all safety networks we have built right now - so scientists are already working on means by which they would be able to protect devices against quantum computers handling cryptography problems.

Another practical application would come from the world of medicine. Classic computers are not very good at doing intricate mathematical operations in chemistry, but quantum computers would be able to handle these without much of a problem. As such, new medicine would be based on the research done with the help of quantum computers.

These are just two of the examples, but there are many, many others. If you want proof that quantum computing is pretty important and that it does mean a lot for the future, just think of the fact that current technology magnates such as Google are redirecting a massive percentage of their funds towards quantum computing research.

If that's not telling enough, we don't know what is!

The Future of Quantum Mechanics: Teleportation?

Last but not least, we want to dabble a little with a concept that is even more fantastic than time traveling, and which has also been discussed for a very long time now: Teleportation.

If time traveling seemed somewhat possible even by the standards of classical physics, teleportation would mean breaking the barrier of everything we have ever known.

How are quantum physics and teleportation related?

In the quantum mechanics point of view, teleportation can be achieved by transporting quantum information from one location to another by using classical means of communication.

The quantum principle lying at the basis of this potential future practice is quantum entanglement. And however crazy it may seem, you might want to learn that scientists have already managed to transport qubits by means of teleportation between two entangled quanta (and they actually managed to do it over pretty impressive distances).

As of recently (in 2019, to be more specific), the teleportation of qutrits has been achieved. For a bit of background, a "qubit" is a piece of quantum information formed out of zeros and ones, while a "qutrit" is a piece of quantum information formed out of zeros, ones, and twos (Nield, 2019).

It has also been reported that scientists have achieved teleportation between two microchips. This might not seem like much, but keep in mind that the transportation of information was not done by regular means (i.e., by simply copying the information from one chip unto the other), but by means of quantum entanglement.

Chapter 3: Quantum Physics - The Battle

Although it is more than one hundred years old, quantum physics is still a relatively new and young science.

As such, proponents of classical physics might simply not agree with many of the theories postulated by quantum physics (and if you look closer at all the arguments currently going back and forth, you might, too, understand why there is a healthy dose of skepticism involved in this entire affair).

This chapter is not meant to ruin everything we have explained thus far, but to show that you should use your own mind to make your own decisions. Unless we are conditioned by quantum entanglement to do certain things because it is "written" (in the stars?), the healthiest thing you can do is try to understand both sides of an argument.

In the world of physics, the main battle is fought between the supporters of Albert Einstein and Erwin Schrödinger, who found some of the theories of quantum physics to be not just unrealistic and improbable, but downright ridiculous at times.

If you want to find out more about the battle between classical physics and quantum mechanics, why we still accept the theories of quantum physics (and why they are still being studied in the most reputable research centers of the world), and what the future of quantum physics might be, then keep reading.

If, however, you prefer to just remain with the theories we have exposed thus far and believe in the amazing perspective of time traveling and bending the barriers of knowledge without hearing "the other side", then you might want to jump to the conclusion section. We do think, however, that it is healthy for you to at least read a bit about the main points behind this debate - so we encourage you to do it even if you are not particularly eager to, and the perspectives shown by quantum mechanics are far more exciting for you (which is completely understandable).

Let's jump in!

Quantum Mechanics and General Relativity Incompatibility

You might have understood this from the book so far, since we have already brushed upon the idea, but quantum mechanics and classical mechanics (as it is seen under the general relativity principle) are quite incompatible at the moment.

Physicists are trying to reconcile the two textbooks according to which science understands the world - but to date, there has surfaced no proven, palpable theory to bring the two worlds together and finally help us understand where we come from, where we are, and where we are going (because, at the end of the day, these are the fundamental questions both classical and quantum physics propose).

In classical physics (as drawn out by Einstein's general relativity principle), reality is made out of 4 dimensions (also called the space-time continuum). In this paradigm, gravitational fields are continuous entities.

In quantum mechanics, however, fields are not continuous, but discontinuous. They are not defined by the 4 dimensions, but by "quanta". As such, concepts like the "gravitational field" are missing from the world of quantum physics, which is also the biggest bridge classical physicists and quantum researchers have to build between their points of view.

This is not just a matter of fancy definitions. The world of quantum mechanics and the world of classical physics are incompatible because they describe reality in completely different ways, in different terms, and in different perspectives that do not meet at any point.

In classical physics, things happen for a reason. They happen according to the old cause-and-effect dictum. Nothing happens randomly, but because there is something else before it that has caused it.

In quantum physics, scientists do not see reality in terms of cause and effect, but in terms of particles jumping from one state to another based on probability, rather than outcomes that are definite.

Why is reconciliation important, then, especially given that these two disciplines seem so different and at such a deep level?

Because reconciliation would also bring along relationships of complementarity. Where classical physics fails to give explanations of the microcosmos, quantum physics would succeed. And where quantum physics fails to make sense when it is blown up to macro objects (remember the cat that was both dead and alive?), classical physics would be able to breathe in some logic.

Einstein's Explanation vs Quantum Mechanics

It is a pretty well-known fact (at least by this point in the book) that Albert Einstein was not a big aficionado of the quantum mechanics theories that were shaping up during his lifetime.

Time proved him wrong in some ways, because some of the quantum theories are actually being proven step by step.

Beyond that, however, the questions posed by Einstein are still valid - and they provide quantum researchers with a point of orientation when it comes to the answers they are yet to give.

If Albert Einstein were alive today, he would probably have "converted" to quantum physics - because even throughout his life, his views on this theory changed. If, at first, his theory completely clashed with quantum mechanics (for the reasons we explained in the previous section of this chapter), he actually used quantum concepts to explain some of his own theories later on in life.

More specifically, in 1935, his experiments revealed what he called "spooky action at a distance" - or, in other terms, quantum entanglement. He then continued his experiments furthering the theory that quantum entanglement was only possible in certain circumstances. Unfortunately, however, he never got a clear answer to this follow-up and it was left to future generations to reconcile the theories.

It would be more than interesting to see what he would have to say about the more recent discoveries and experiments in quantum mechanics.

Why Do We Accept Quantum Mechanics?

Without a doubt, Einstein's work reshaped the world in so many ways that it would take an entire library of books to simply explain them in plain English. In the scientific community (and, dare we say, outside of it too), Einstein is seen as a sort of demi-god - an irrefutable authority that nobody dares to touch.

Nobody except quantum physicists, that is.

If Einstein's theory of relativity is so well-regarded and accepted, why do we even bother with quantum mechanics, then? What demon sets so many contemporary scientific researchers on the path of actually trying to reconcile the worlds of classical physics and quantum physics?

Well, the one reason quantum mechanics is accepted and still very much a topic of discussion is because it would actually manage to solve what classical physics couldn't. And, as it has been shown throughout this book, it would actually manage to push the boundaries of knowledge and technology beyond the edges of the imaginary and into a spectrum we only dared to touch with our thoughts until not very long ago.

September 7, 2014 might have seemed like any other day of fall in the Northern Hemisphere. The leaves were probably slightly yellow by then and the heat of the summer was slowly starting to wear out. Maybe it even rained a little in the morning, and by the time cities were waking up to life, the fog of a slightly chillier night vanished, leaving room for a perfect day of autumn.

What everyone must know is that September 7, 2014 was the day the Theory of Everything officially saw the light of day. You might have heard about it because there was a movie on the life of Stephen Hawking. Or you might have even stumbled upon it long before the movie came out.

What is important, however, is that the Theory of Everything is one of the most important attempts at unifying both the theory of relativity and the quantum theory. What was started in the 1920s by Albert Einstein was finally starting to make sense eight decades later under the hands of Stephen Hawking.

The Theory of Everything is, perhaps, one of the most ambitious projects ever. It is one of the theories that is bound to change every single little thing - not just in physics, but in science as a whole, and, soon enough, in mankind's perception of pretty much every area of their lives.

What the Theory of Everything tries to do is finally build a bridge between quantum mechanics and the theory of relativity. Some would even dare to say that it will "tell the mind of God" (Marshall, 2010) and that it will hold the key to mankind answering the questions it has been trying to answer for a very, very long time now.

There are several candidates for the Theory of Everything. Some of them are implausible to be actually proven in equation or in practice, but some of them stand out as sane options that might be the final answer to everything.

Out of these, we would like to take the time to name the two most important contenders. As we draw close to the conclusion of this book, we believe it is important for you to know what the most important work in physics is doing now - and as such, we will take the time to expand, just a little bit, on these two theories.

One of them is called "String Theory", and what it says is that there is a ten-dimensional space we are living in. That sounds more than mind-boggling, we know, but wait until you hear more of it.

According to the String Theory, the point-like particles of particle physics are one-dimensional objects (called "strings"). The theory describes that these strings propagate through space and that they interact with each other. When looked at from afar, a string looks like any other ordinary particle (with a mass, charge, etc. that are determined by the vibrational state of the string). For instance, one of the vibrational states of the string is represented by gravitation (a particle that carries the gravitational force, that is).

In essence, the Theory of Everything relies on quantum gravity and it aims to address a wide range of questions in fundamental physics - such as what is going on with black holes, how the universe was formed, how to improve nuclear physics, and how to handle condensed matter physics better.

Ideally, string theory will unify gravity and particle physics (which is one of the main points that have to be bridged between classical physics and quantum mechanics). At the moment, however, it is not clear how much of this theory can be adapted to the real world and how much of it will allow for changes in its details.

The other theory competing with string theory for the title of "The Theory of Everything" is the

Loop Quantum Gravity Theory. This paradigm is heavily based on Einstein's work, and it was elaborated towards the middle of the 1980s. To understand it, you need to remember the fact that, according to Einstein, gravity is not a force per se, but a property of space-time.

Up until the Loop Quantum Gravity Theory, there have been several attempts to prove that gravity can be treated as a quantum force, like electromagnetism or the nuclear force, for example. However, these attempts have failed.

What the Loop Quantum Gravity Theory tries to do is to base the bridge between traditional physics and quantum physics on Einstein's geometric formulation. Ideally, this will prove that space and time are quantized the same way as energy and momentum are in quantum mechanics.

If physicists manage to prove the Loop Quantum Gravity Theory, space-time will be pictured with space and time being granular, which would consequently mean that a minimum space exists. In other words, according to the Loop Quantum Gravity

Theory, space is made out of a fine fabric of woven finite loops called "spin networks".

Although String Theory seems to be a lot more popular in mainstream media (mainly due to the fact that some of its proponents are quite popular themselves, even well outside of scientific circles - like Michio Kaku, for example), the Loop Quantum Gravity Theory should not be dismissed in any way. Most of its implications are related to the birth of the universe, reason for which it is also called the Big Bang Theory - and, perhaps, the reason for which the eponymous TV show was called that way as well.

In addition to the string and loop quantum gravity theories, you might also stumble upon a number of other candidates to become the Theory of Everything. Some of them include the Causal Dynamical Triangulations Theory, the Quantum Einstein Gravity Theory, the Quantum Graphity Theory, or the Internal Relativity Theory.

All of these theories show that active efforts are being put into unifying quantum theory and the more classical physics, proving that the vast majority of the science community pays quite a lot of attention to quantum mechanics.

Who are we to dismiss them, then? Just because things are still foggy, it doesn't mean that they will stay forever this way. And, at the end of the day, the whole essence of science in general is to *dream* and *aim for* something bigger, more comprehensive, and more efficient. It has always been this way, and it will always be.

And when it comes to the *ultimate dream*, nothing gets as close to the grandeur, the superbity, and the brevity of the quantum mechanics world - precisely because it is the only theory that will finally give us a well-deserved push forward into a wide range of discoveries.

What should *you* believe?

It is up to you. If this book has made you curious, then we invite you to learn more about both quantum and classical physics and to make up your own mind. The beauty about physics and research in this field is that nothing is ever fixed, and that theories that might have seemed unbreakable have been consistently broken over the course of history - starting with, for example, the flatness of the Earth.

Believe what you think is true based on your own readings and research, but stay true to the fallible nature of everything!

Conclusion

"Look again at that dot. That's here. That's home. That's us. On it, everyone you love, everyone you know, everyone you ever heard of, every human being who ever was, lived out their lives. The aggregate of our joy and suffering, thousands of confident religions, ideologies, and economic doctrines, every hunter and forager, every hero and coward, every creator and destroyer of civilization, every king and peasant, every young couple in love, every mother and father, hopeful child, inventor and explorer, every teacher of morals, every corrupt politician, every 'superstar,' every 'supreme leader,' every saint and sinner in the history of our species lived there--on a mote of dust suspended in a sunbeam." (Sagan, 2011)

The thing with life and everything we know is that they are inexplicably intricate, weird, and a lot of the time, paradoxical. The very fact that we are born to die, rather than live (in most people's acceptation of life) is mind-boggling in itself. Yes, one might argue that between birth and death, we are filled with moments that make life meaningful - and we absolutely agree with that.

Yet, if you take all that out of the equation, so to speak, you will very easily come to the conclusion that the final destination of birth is death. It is the paradox of being alive, if you want to put it this way.

For centuries, science has postulated unbeatable truths. And for centuries, the same unbeatable truths have been ruthlessly demolished by the very guild that had elevated them to the state of absolute truths.

They said the Earth is flat, but then, they realized it isn't.

They said we cannot fly, but then, before we even realized what was going on, we were flying low-cost to see grandma over Christmas.

They said the moon landing is a crazy dream, but then, we actually landed on the moon before we even started to fly low-cost.

The world of science is intrinsically fascinating when you look at it because it always seems that there is a battle between one clique and another one. A mental battle, of course - one in which formulae are tossed at the adversary team and theories are postulated as if they were firing guns to "kill" the other team's ideas.

The battle between quantum physics and classical physics is nothing new, really, because for every major discovery mankind has ever made, there have always been a bunch of skeptics who said that it cannot be done (or that it should be done in a far more complicated way).

If the history of science has proven anything, it is that the more we step forward, the more questions we have, and the more ardent the battle between the two opposing forces of science is (at one time or another in history).

In fact, the world of science does nothing more than mirror the way in which we are built, the paradoxes upon which we have stepped down from the trees and started making fires and growing our own food.

You see, even if you look beyond the "live to die" paradox of human life, we, humans, are flawed by a million and one paradoxes. We all aim for peace because war has proven to be absolutely atrocious - and yet when the time comes, we'll be ready to hit our neighbor with a stone. We aim for technical advances - and yet, when something new comes along, we run scared and start bringing arguments against those technical advances.

Science does nothing else than to mirror our state - on the one hand anchored in tradition, on the other hand permanently disrupting itself from the boundaries of tradition, science has never been as confusing, as exciting, and as mind-bogglingly beautiful as it is today.

It may be because we have more scientific creative freedom than ever, with secularity now being a fact for the vast majority of the world.

It may be because we have finally reached a stage where we not only want to discover more than just the obvious, but actually move past it and discover the less obvious - the truths that have been transpiring through nature, but which we were too scared, or, if we have to be honest, too silly to reach out for.

Without a doubt, though, we are living in times of extreme change - and quantum physics is right there, as the harbinger of all that is to come. We may not fully understand it, like children who are just learning how to distinguish colors and make experiments with crayons on paper. But it will not be long before we will be able to fully grasp the meaning of all this quantum gibberish.

And when that happens, well, we're bound to reach the stars.

Yes, the actual stars.

There is a world of knowledge out there, within and around the stars, beyond what we have been able to observe from the pale blue dot until now. To many people, it might actually seem too daring to think that we might one day be able to reach the stars and live a life torn out of science fiction scenarios.

To those who are religious, knowing the absolute essence of nature and the world might actually seem like too much, as if we are getting too close to God and his own nature, in a way.

And to those who are not religious, but simply skeptical by nature, quantum physics might all seem like just plain gibberish talk - the kind they do on TV when there is nothing really important to discuss.

If you have gone through this book, it means that you were at least curious to discover the world out there and the theories in physics that are trying to describe it through a modern lens - a lens that might seem broken and that might send out a disparate imagery, but one that, we'd have to admit, is one of the single most interesting ones we have ever looked at.

If you are at this point in the book, we cannot but congratulate you. We know quantum physics can be a mind-boggling subject, especially when you try to pick it up in a disparate and chaotic way. It is a field that is still in the works, which makes it highly confusing for any layman, no matter their background - and, if we have to be honest, makes it highly confusing even for people within the realm of physics in general.

Going through this book means that you are highly interested in discovering the secrets of the world - and that is a major, major quality in anyone. Regardless of whether you are teaching history or math, if you work in a factory or in retail, if you have a boring office job or are out there on the field saving lives, quantum physics is a subject that will interest you because it is very likely that your own field of expertise will be affected by it one day.

To say that quantum physics is not realistic or that it has no real applications is the same as saying that cloning is not possible and that it has absolutely no value (outside of the metaphysical and ethical implications of it).

To say that you don't want to be interested in the basics of quantum physics at least, is to say that you are not interested to know where you came from and where you are going as a member of this massive family called "mankind".

Yes, we genuinely believe quantum mechanics is the key to prevailing and to pushing our existence one step forward in so many ways. It's not about the SF things only, like time traveling or teleportation (although, let's admit it, those would be terribly nice to have, wouldn't they?). It's about medicine, energy, air pollution, the world as a whole, and the comfort of 7 billion of us, crammed into a planet that is nothing but a speck of dust in a tremendously large universe.

It is impossible not to have at least the vaguest, most remote interest in quantum mechanics. Look at it this way: Even if you care zero about mankind and its advancements, you still care about where you will be two or three decades from now. For instance, if you are a driver, you don't have to be directly interested in the mechanics and electronics of how your car is running, but you do have to be interested in the basics behind it because you might find yourself with a broken car in the middle of nowhere at one point or another.

The same goes with quantum physics as well: You don't have to be interested in the intricacies of the ballet that the equations form on every physician's whiteboard right now. But you should be at least minimally interested in what is going on, precisely because all of this is bound to be the future of our species (and yes, you *are* part of that future as well, especially given how fast things are advancing in this field).

Quantum physics is the key to unleashing the far-fetched imaginary future we have always written and sang about: A future where nothing is a boundary anymore, where medical advances happen every day, where humans can travel freely into space and time, where we are limited by absolutely nothing.

Being at least vaguely familiar with the main theories of quantum physics means you are in the know with what the future is brewing - and we genuinely hope our book has helped you move one step closer to that.

We strongly encourage you to read more and discover more about this topic, if not for anything else, then at least for your mental exercises.

Good luck in your further reading and researching, and don't forget to dream about the impossible, because science might just make it come true!

References

Ball, P. (2019). Quantum Leaps, Long Assumed to Be Instantaneous, Take Time | Quanta Magazine. Retrieved 3 January 2020, from https://www.quantamagazine.org/quantum-leaps-long-assumed-to-be-instantaneous-take-time-20190605/

Greene, T. (2019). [Best of 2019] This quantum physics breakthrough could be the origin story for time travel. Retrieved 3 January 2020, from https://thenextweb.com/science/2019/10/03/this-quantum-physics-breakthrough-could-be-the-origin-story-for-time-travel/

MacDonald, F. (2019). Scientists Just Unveiled The First-Ever Photo of Quantum Entanglement. Retrieved 3 January 2020, from https://www.sciencealert.com/scientists-just-unveiled-the-first-ever-photo-of-quantum-entanglement

Marshall, M. (2010). Knowing the mind of God: Seven theories of everything. Retrieved 3 January 2020, from https://www.newscientist.com/article/dn18612-knowing-the-mind-of-god-seven-theories-of-everything/

Nield, D. (2019). Quantum Teleportation Has Been Reported in a Qutrit For The First Time. Retrieved 3 January 2020, from https://www.sciencealert.com/quantum-teleportation-has-been-reported-in-a-qutrit-for-the-first-time

PhotonTerrace. (2020). History of research on light | Photon terrace. Retrieved 3 January 2020, from https://photonterrace.net/en/photon/history/

Sagan, C., & Druyan, A. (2011). *Pale blue dot*. New York: Ballantine Books, an imprint of The Random House Publishing Group, a division of Random House.

Wikipedia. (2020). Quantum mechanics. Retrieved 3 January 2020, from https://en.wikipedia.org/wiki/Quantum_mechanics#History

Wikipedia. (2020). Superposition principle. Retrieved 3 January 2020, from https://en.wikipedia.org/wiki/Superposition_principle

Printed in Great Britain
by Amazon